First published in Great Britain 2023 by Farshore
An imprint of HarperCollins*Publishers*
1 London Bridge Street, London SE1 9GF
www.farshore.co.uk

HarperCollins*Publishers*
Macken House, 39/40 Mayor Street Upper, Dublin 1, D01 C9W8, Ireland

Written by Claire Philip
Images used under license from Shutterstock.com

© The Trustees of the Natural History Museum, London 2023

ISBN 978 0 00 859114 4
Printed and bound in UAE
001

Parental guidance is advised for all craft and colouring activities.
Always ask an adult to help when using glue, paint and scissors.
Wear protective clothing and cover surfaces to avoid staining.

A CIP catalogue record for this title is available from the British Library.

Farshore is not responsible for content hosted by third parties.

Farshore takes its responsibility to the plant and its inhabitants very
seriously. We aim to use papers from well-managed forests run by
responsible suppliers.

DINOSAURS

ANNUAL 2024

CONTENTS

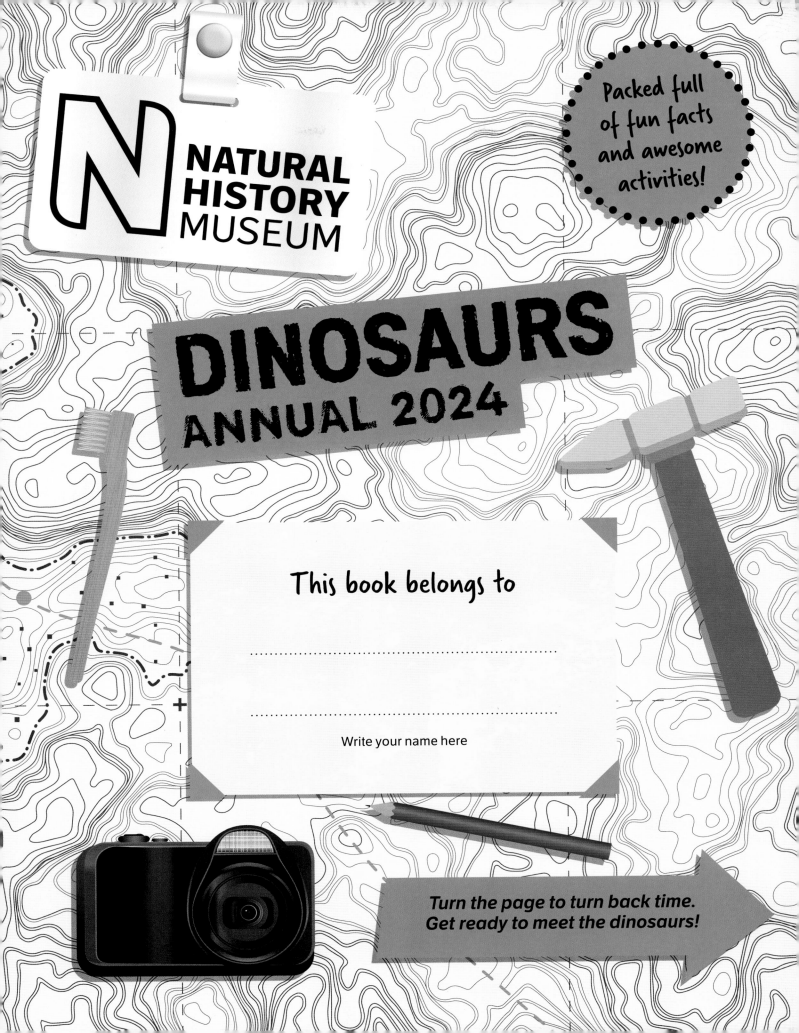

NATURAL HISTORY MUSEUM

Packed full of fun facts and awesome activities!

DINOSAURS
ANNUAL 2024

This book belongs to

....

....

Write your name here

Turn the page to turn back time. Get ready to meet the dinosaurs!

LOOKING INTO THE PAST

Did you know that the Earth's continents didn't always look like they do today?

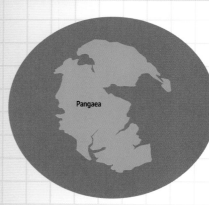

Permian Period
299-251 Million Years Ago

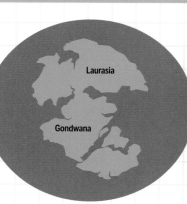

Triassic Period
252–201 Million Years Ago

Jurassic Period
201–145 Million Years Ago

Earth's outer layer is made up of several enormous moveable pieces called tectonic plates, which fit together like a jigsaw puzzle.

Cretaceous Period
145–66 Million Years Ago

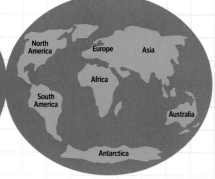

Present day

Tectonic Plates

Over the millions of years that dinosaurs were alive, Earth's tectonic plates shifted and the land mass of Pangaea broke up into separate parts. The dinosaurs on the different parts adapted to their changing environments by developing unique characteristics that helped them to survive.

Amazing Evolution

Due to the changing conditions on Earth, life on our planet has evolved from very simple life forms all the way to complex human beings!

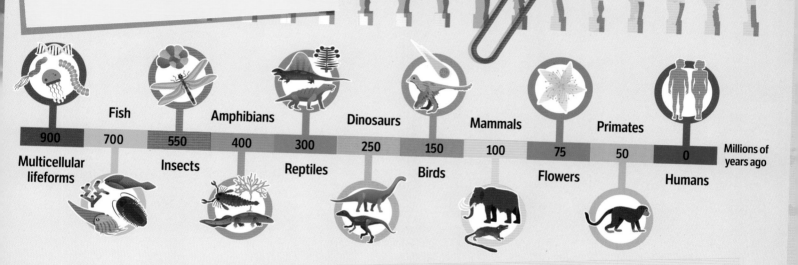

900	700	550	400	300	250	150	100	75	50	0

Fish — Amphibians — Dinosaurs — Mammals — Primates

Multicellular lifeforms — Insects — Reptiles — Birds — Flowers — Humans

Millions of years ago

Dinosaur Timeline

The dinosaurs lived for about 180 million years over three periods of Earth's history. These are called the Triassic, Jurassic and Cretaceous Periods.

The **first dinosaurs** evolved from other reptiles after a large extinction event caused most of the other species to die out. Herrerasaurus was one of the earliest dinosaurs.

Near the end of the **Triassic Period** (252–201 million years ago), huge volcanic eruptions and earthquakes took place. This caused Earth's giant land mass, Pangaea, to split.

During the **Jurassic Period** (201–145 million years ago), plant life flourished. This led to huge plant-eating dinosaurs such as Apatosaurus and Brachiosaurus.

The continents spread apart even more throughout the **Cretaceous Period** (145–66 million years ago). This meant that many more dinosaur species evolved. Meat-eating Spinosaurus roamed at this time.

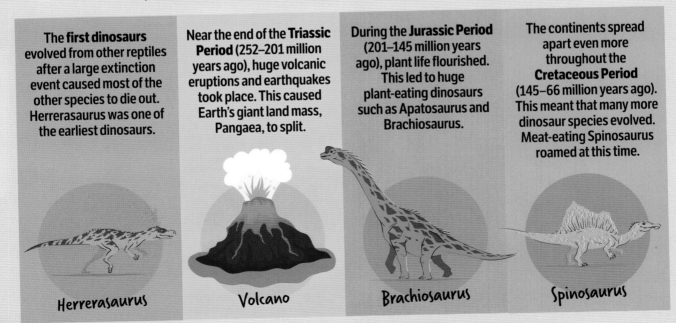

Herrerasaurus — Volcano — Brachiosaurus — Spinosaurus

The dinosaurs died out around 66 million years ago.

STYRACOSAURUS

A super spikey plant-eater!

STOMP!

Horned Dinosaur

Styracosaurus belonged to the horned dinosaur group known as the ceratopsians. It roamed Earth during the late Cretaceous period. This plant-eater was related to the famous dinosaur Triceratops, though it was much smaller. Most fossils of Styracosaurus have been found in Alberta, Canada.

Short tail

DINO FACTS

Name: *Styracosaurus*

Meaning: *"Spiked lizard"*

Size comparison:

Food: *Plants*

Danger rating: *4/10*

Where: *Coastal plains and woodlands*

Safety in Numbers

It is thought that these dinosaurs may have lived together for safety – fossil evidence suggests that they may have lived in herds. Their remains are often found in groups. At one fossil site there were more than 100 individuals.

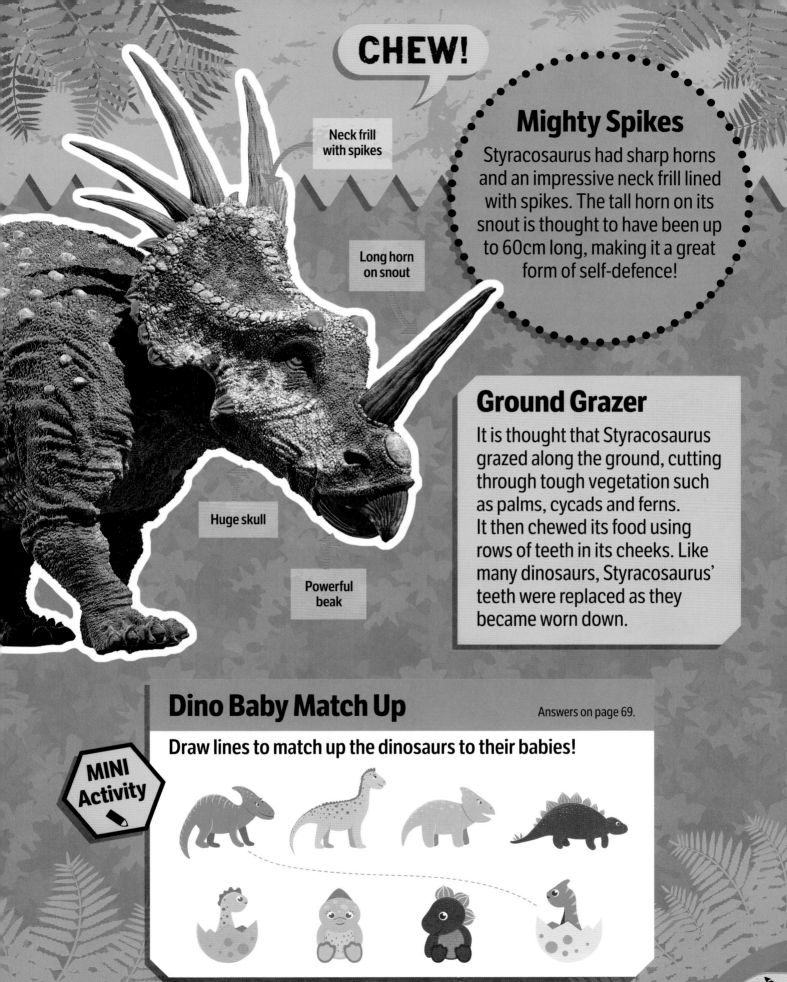

CHEW!

Neck frill with spikes

Long horn on snout

Mighty Spikes

Styracosaurus had sharp horns and an impressive neck frill lined with spikes. The tall horn on its snout is thought to have been up to 60cm long, making it a great form of self-defence!

Huge skull

Powerful beak

Ground Grazer

It is thought that Styracosaurus grazed along the ground, cutting through tough vegetation such as palms, cycads and ferns. It then chewed its food using rows of teeth in its cheeks. Like many dinosaurs, Styracosaurus' teeth were replaced as they became worn down.

Dino Baby Match Up

Answers on page 69.

Draw lines to match up the dinosaurs to their babies!

MINI Activity

MAZE CHALLENGE

This flying reptile has lost its way while hunting for food.
Can you find the path back to its hungry babies?

START

FINISH

Answers on page 69.

COLOUR BY NUMBERS

This Stegosaurus needs colouring in!
Use the colour key below to complete the picture.

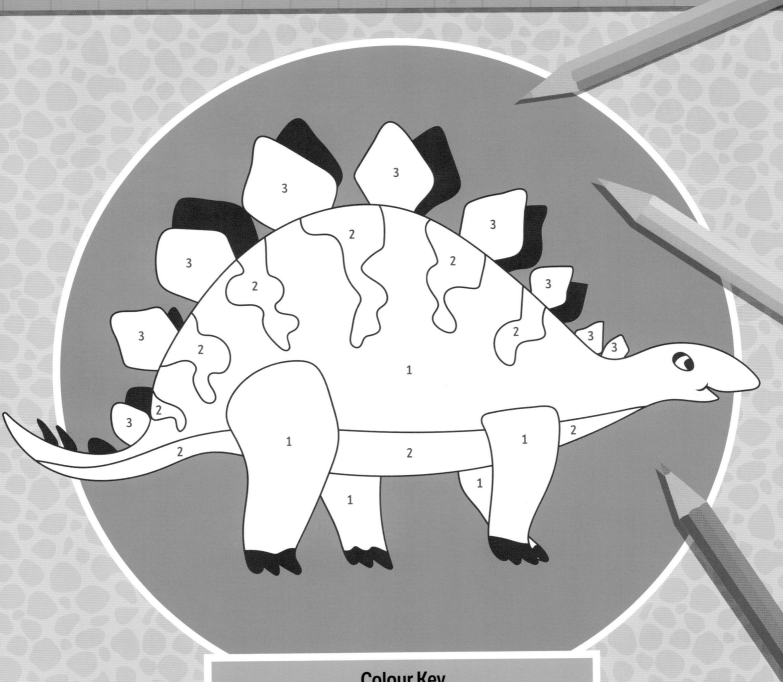

Colour Key

1 2 3

13

IGUANODON

This famous dinosaur was discovered in the south of England in 1822 – that's just over 200 years ago.

Body Features

Ten-metre-long Iguanodon was a plant-eating dinosaur from the Cretaceous period. It is thought that it spent most of its time walking on all fours, but that it could make speedy getaways by running on its back two legs.

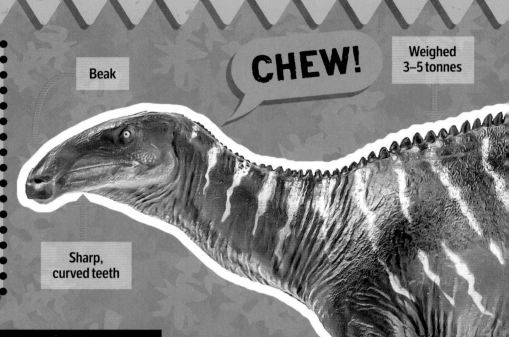

Beak

CHEW!

Weighed 3–5 tonnes

Sharp, curved teeth

Thumb spikes on both hands

DINO FACTS

Name: *Iguanodon*

Meaning: *"Iguana tooth"*

Size comparison:

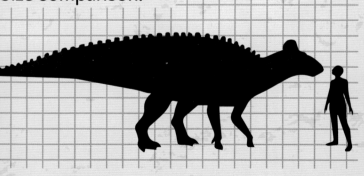

Food: *Plants*

Danger rating: *3/10*

Where: *Forests, plains, swamps*

Deadly Spikes

On its hands were razor-sharp thumb spikes. Experts think these were used for self-defence, or perhaps to slash through thick vegetation. Watch out – coming through!

SLASH!

14

Fossil Finds

In 1878 an amazing thirty-eight Iguanodon skeletons were discovered in a Belgian coal mine. Perhaps, like many other herbivores, this dinosaur lived in groups for safety.

Long, strong tail

Mirror Image

Use the grid to copy the image. Colour it in to finish the picture.

Early Discovery

Gideon Mantell gave this dinosaur the name "Iguanodon" because he thought its teeth looked like those of a modern iguana! It was only the second species to be scientifically named a dinosaur – the first was Megalosaurus.

ROLL THE DICE

All you need to play this board game is a friend, two counters and a dice.

To decide who goes first, roll the dice. The highest number wins.

To play, take it in turns to roll the dice. Move your counter the number of spaces shown.

If you land on the green arrows you get to skip ahead. Bad luck if you land on a red arrow, you'll have to go back!

Who will get to the finish line first?

Colour in the dino friends.

WHAT IS A DINOSAUR?

When dinosaur fossils were first discovered, nobody knew what they were!

A Brand-New Group

It wasn't until the 1800s that Sir Richard Owen used the word "dinosaur" to describe a group of fossils that he had been studying. Before that, people thought dinosaur remains belonged to other large reptiles — or even giant mammals.

How are Dinosaurs Classified?

All dinosaurs are divided into two main groups: dinosaurs with lizard-like hips and dinosaurs with bird-like hips. Unlike lizards, whose legs sprawl out from their hips, dinosaurs had an upright stance with their legs directly under their bodies.

Dinosaur Knowledge

We know so much about dinosaurs — such as how they lived, what they ate and how they moved — from fossil evidence. Fossils are the preserved rocky remains of an animal or plant. Around the world, people have found all kinds of dinosaur fossils from teeth and bones to footprints, eggs and even poo!

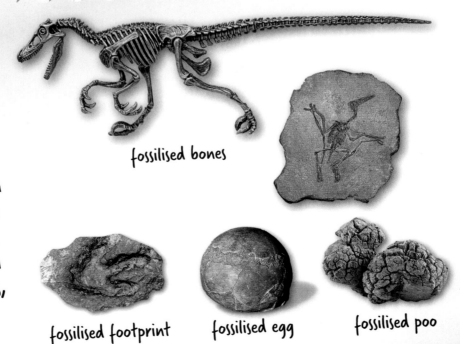

fossilised bones

fossilised footprint

fossilised egg

fossilised poo

Glossary Box:

Bird-hipped dinosaurs are also called **ornithischian** dinosaurs.

Lizard-hipped dinosaurs are also called **saurischian** dinosaurs.

An **asteroid** is a huge space rock that travels around the Sun.

What Happened to the Dinosaurs?

Experts believe that 66 million years ago, a giant asteroid crashed into the surface of our planet, landing in Mexico. The crater formed on impact was huge — 150 kilometres from one side to the other! The crash caused lots of dust to fly up into the air, blocking out a huge amount of sunlight. It is thought that around three-quarters of all animals died out along with the dinosaurs.

Do Dinosaurs Still Exist?

Not like they did millions of years ago! However, some of their distant relatives do still live among us. Birds evolved from the theropod dinosaur group that included Velociraptors. Like birds, theropods walked upright on two legs and some had feathers.

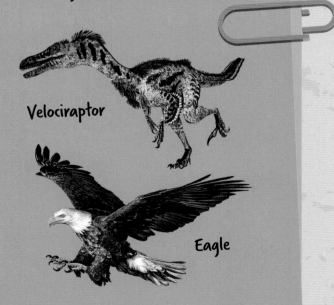

Velociraptor

Eagle

PLESIOSAURUS

This marine reptile lived alongside dinosaurs during the early Jurassic era!

GLIDE!

REPTILE FACTS

Name: *Plesiosaurus*

Meaning: *"Close to lizard"*

Size comparison:

Food: *Fish and other sea creatures*

Danger rating: *4/10*

Where: *Water*

Fish Eater

This ocean dweller spent its time searching for fish and other sea creatures to munch! It is thought that its sharp, cone-shaped teeth and strong jaws allowed it to eat hard and crunchy molluscs such as clams.

Broad tail

Narrow fins

The First Specimen

In 1823, fossil hunter Mary Anning discovered the first-ever complete Plesiosaurus skeleton, but other paleontologists thought that the fossil was a fake! After meeting to discuss it, the scientists eventually agreed that it was real. In 1824, the highly respected paleontologist Georges Cuvier said: "It is the most amazing creature ever discovered".

SPLASH!

Small head

Long neck

Turtle-like body

Around 4.5 metres long

Come Up for Air

Plesiosaurus had to come to the surface to breathe air as it had lungs (like land-dwelling reptiles) instead of gills (like fish). We don't know how long it was able to stay underwater without breathing.

MINI Activity

Counting Creatures

Colour in the boxes on the graph for each creature you count.

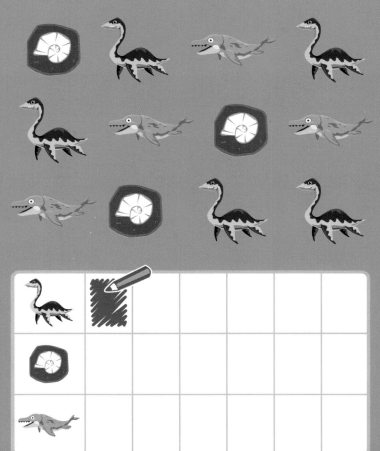

Fantastic Fins

The four long fins of Plesiosaurus were perfectly adapted to its watery environment. They helped this marine reptile glide through the ocean in a smooth flying motion, a bit like a modern-day turtle.

Answers on page 69.

LET'S COLOUR!

Colour in the picture of the Ankylosaurus using the finished picture as a guide.

22

COLOUR BY NUMBERS

This flying reptile needs colouring in!
Use the key to complete the picture.

1
2
3
4
5
6

MEGALOSAURUS

The first prehistoric creature to be called a dinosaur!

The First Dinosaur!

In the 1820s, Megalosaurus was the first prehistoric creature to be called a dinosaur. It was identified by several fossils, including a piece of lower jaw lined with sharp teeth. A model of Megalosaurus was made based on the early fossil findings and put on display to show the public.

Long tail held off the ground

Strong back legs

DINO FACTS

Name: *Megalosaurus*

Meaning: *"Great lizard"*

Size comparison:

Food: *Meat*

Danger rating: *6/10*

Where: *Tropical forests*

Theropod Family

A complete skeleton of Megalosaurus has never been found, so we don't know exactly what it looked like. However, it is thought to have walked upright like other theropod family members, including T.rex and Allosaurus. It grew up to 3 metres tall and weighed around one tonne. Wow!

STOMP!

CHOMP!

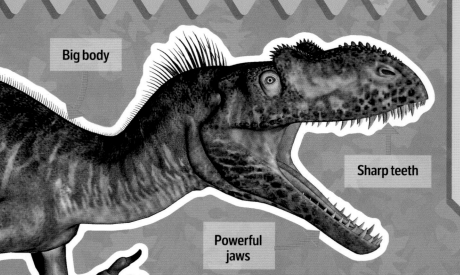

Big body

Sharp teeth

Powerful jaws

Short front arms

Deadly Hunter

Megalosaurus was a carnivore and probably hunted large, plant-eating dinosaurs called sauropods. Its ideal habitat was a lush tropical forest, as there would have been plenty of grazing prey to hunt and eat – and plenty of places to hide!

MINI Activity

Prehistoric Puzzle

Complete the grid by drawing the correct picture in each of the empty spaces. Each object should only appear once in each box, column and row.

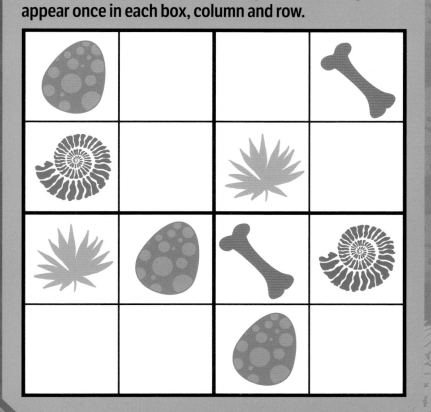

Answers on page 69.

SPOT THE DIFFERENCE

How quickly can you spot the ten differences between the two pictures?

A

Start a stopwatch and see how long it takes you.

Ask a friend to have a go – who was the quickest?

It took me [ ] minutes and [ ] seconds to find all ten differences.

Circle the differences as you find them. Colour in the dinosaur eggs as you go!

1 2 3 4 5 6 7 8 9 10

Answers on page 69.

FROM BONES TO STONES

Dinosaur fossils have been found all around the world, even in Antarctica!
Here are some of the most important sites.

Fossil Formation

After an animal dies, most of its body rots away. Only the hard skeleton and teeth are left behind. Over time, and if the animal is close to water, these remains are covered by layers of mud and sand. Over millions of years, this forms into sedimentary rock. During this process, the bones and teeth are turned into stone by minerals in the water. Over time the rock layers wear away exposing the fossils for us to discover!

North America

This T.rex fossil is one of the most complete ever found and is nicknamed 'Sue'

Europe

Fossilised claw of a Baryonyx (fish eating dinosaur) found in the UK

South America

These fossilised eggs of a Neuquensaurus were found in Argentina

Learn the Lingo!

★ Body fossils include shells, bones, claws and teeth.

★ Mould fossils are the impressions of an animal or plant left behind in rock.

★ Cast fossils are formed when these impressions are filled in by minerals.

★ Trace fossils are remains of an animal such as their poo or imprints of their footprints, feathers or skin.

When fossil hunting, always check if you are allowed to remove any fossils you find and always go with an adult. If you find something unusual or impressive, take it to a local museum.

28

How to Find Fossils

Ask an adult to help you find somewhere well known for fossil discoveries. Rocky beaches can be a great place to start! Only go when the tide is going out so that the shore isn't covered by the sea. You'll need a notebook and pencil to make notes and a small bag to place any findings in. You may wish to take a camera, too.

The Jurassic coast, Dorset, England.

Glossary box:

Sedimentary rock is made from compacted material such as sand, mud, minerals, fossils and plants.

A **paleontologist** is a scientist who studies animal and plant fossils.

Asia

Keichousaurus hui fossil, a marine reptile found in China

Australia

Dinosaur footprint found in Western Australia

Antarctica

This is an illustration of a Cryolophosaurus skull, a Jurassic dinosaur discovered in Antarctica

Who Was Mary Anning?

Born in 1799, Mary Anning was a fossil collector from southwest England. She made some incredible fossil discoveries, including two marine reptiles — an ichthyosaur and a plesiosaur, shown here displayed at the Natural History Museum — and a type of pterosaur called a Dimorphodon.
During her life, her work wasn't taken very seriously but now she is recognised as an important paleontologist.

DIPLODOCUS

A large, long-necked plant eater from the late Jurassic era!

Long Bodies

One of the longest dinosaurs of all time, Diplodocus could reach an amazing 30 metres from the tip of its nose to the end of its tail! Most of its length was made up of its neck and tail, which had two rows of bones to give it enough support. The tail could be used in self-defence, and the neck allowed the dinosaur to reach high-growing leaves.

Tiny brain

Very small head for body size

Very long body

CHEW!

DINO FACTS

Name: *Diplodocus*

Meaning: *"Double beam"*

Size comparison:

Food: *Plants*

Danger rating: *4/10*

Where: *Plains and woodlands*

A Rocky Meal

Diplodocus had small, nibbling teeth at the front of its mouth that were replaced when they became worn down. It is thought that it couldn't chew its food so it may have needed to swallow rocks or pebbles to help grind up the plant matter when it reached its stomach.

Finding Friends

Answers on page 69.

MINI Activity

1
2
3

Friends

Trace the wiggly lines to find which path the dinosaur should take to get to its friends!

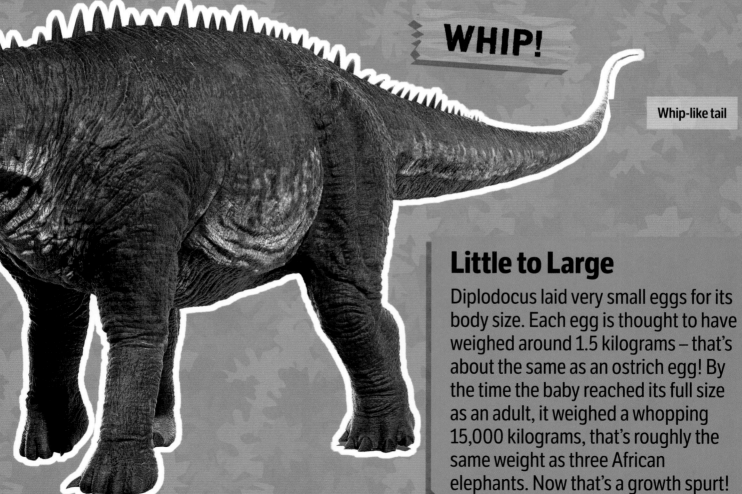

WHIP!

Whip-like tail

Little to Large

Diplodocus laid very small eggs for its body size. Each egg is thought to have weighed around 1.5 kilograms – that's about the same as an ostrich egg! By the time the baby reached its full size as an adult, it weighed a whopping 15,000 kilograms, that's roughly the same weight as three African elephants. Now that's a growth spurt!

A claw on both front feet

DINO DOUGH

Here's a fun rainy day activity using play clay or dough.

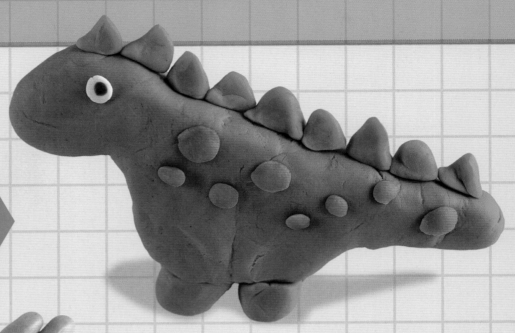

Ask an adult to help !

TIP:
Make sure you cover any surfaces before you start!

You will need:

- Clay or dough in three colours, one larger piece and two smaller pieces
- Googly eyes, or some white and black dough
- Protective mat

Step 1

Take a small piece of the larger ball of dough and set aside. Shape the rest into a flat triangle with one long edge. Form one end into a round 'head' shape, and the other end into a longer 'tail' shape. Take the small piece you set aside and roll into four small legs and attach as shown.

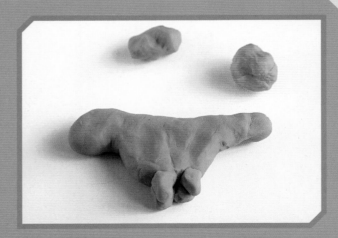

It should look like this picture.

Step 2

Take one of the small balls of dough, pinch into small triangle shapes and attach along the back of your dinosaur.

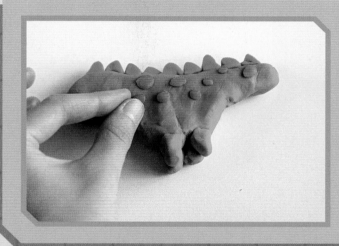

Step 3

Using the last ball of dough, roll small circles and press into the body of your dinosaur. You could also make stripes or feathers!

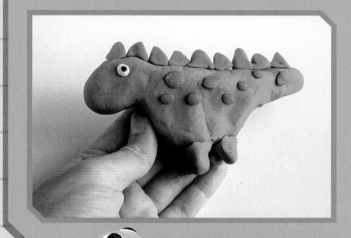

Step 4

Finally, finish off your dinosaur creation with small google eyes, or some white and black dough.
Roar!!

TIP:

Once you have made your roarsome dough dinosaur, why don't you make some more? Try experimenting with different colours and patterns. Here are some more examples.

TYRANNOSAURUS REX

The ultimate meat-eater from the late Cretaceous period!

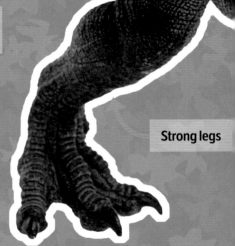

Speedy Killer

It is thought that T.rex's strong legs and upright posture allowed it to run at speeds of up to 20 kilometres an hour. This ability to chase, along with its sharp teeth and powerful jaws, made it an expert hunter.

Long tail

Up to 6 metres tall

Strong legs

DINO FACTS

Name: *Tyrannosaurus rex*

Meaning: *"Tyrant lizard"*

Size comparison:

Food: *Meat*

Danger rating: *10/10*

Where: *Forests and swamps*

Strength in Numbers

Fossils from what looks like a family of T.rex have been found, suggesting that these huge meat-eaters may have lived and hunted in packs – like wolves! It mostly ate plant-eating dinosaurs, such as Triceratops, but it probably scavenged meat from dead animals, too.

34

SLICE!

Large brain

ROAR!

Banana-sized teeth

Small arms

Little Arms

T.rex is known for having small arms compared to its large body size. Experts aren't exactly sure why they are so tiny. One recent theory is that the adaptation may have helped it avoid injury from other T.rex during mass feeding frenzies!

MINI Activity

Shadow Match

Only one of these shadows matches this picture of a T.rex, can you spot which one?

1

2

3

4

WEIRD, COOL AND GROSS!

YUK!

1 The biggest fossilised dinosaur poo ever found weighs nearly 10 kilograms. That's the same weight as a car tyre!

2 Scientists think that dinosaurs could catch colds! Imagine the sneezing – and the snot!

WEIRD!

3 Giant fleas fed on dinosaur blood! They had sucking tubes to pierce their skin.

5 When born, a baby T.rex was about the same size as a turkey!

4 Microraptor was one of the smallest dinosaurs. It weighed less than one kilogram!

CUTE!

Impress your family and friends with these diverting dino facts.

COOL!

6 The spikes at the end of some dinosaur tails have an awesome name – a thagomizer!

8 It is thought that some dinosaur species ate each other as well as their prey!

GROSS!

7 Just like animals today, dinosaurs would have probably burped and farted!

WOW!

10 So far, the longest dinosaur name is Micropachycephalosaurus. It means "tiny, thick-headed lizard." Can you say it? "my-cro-pack-ee-seff-allo-saw-rus"

9 Dinosaurs may have shed their scales – imagine the dandruff!

HA! HA!

FACT FILE
OVER AND UNDER

Meet the prehistoric reptiles that lived in the sea and took to the skies.

Sea and Sky

Even though some dinosaurs could swim and some had feathers, they all lived on the land. The creatures on this page are often thought to be dinosaurs, but they belong to separate animal groups.

Pteranodon

Pteranodon had long jaws but no teeth. It is thought that this super flier caught and ate fish like a pelican.

Plesiosaurs

Plesiosaurs were giant marine reptiles that had either long or short necks and paddle-like flippers. They lived from the late Triassic to the late Cretaceous period.

Eudimorphodon
Eudimorphodon was one of the first pterosaurs to evolve and was quite small.

Quetzalcoatlus northropi
The largest pterosaur – called **Quetzalcoatlus northropi** – was as big as a plane!

Mosasaurs
Mosasaurs were lizard-like swimmers that lived during the Cretaceous period. They died out at the same time as the dinosaurs.

Ichthyosaurs
Ichthyosaurs were sharp-toothed ocean predators that lived during the Jurassic era.

UTAHRAPTOR
The biggest raptor to walk the Earth!

ATTACK!

In the Family

Utahraptor was a dromaeosaur, a type of dinosaur belonging to the bird-like therapod group. Dromaeosaurs ranged in size, but all were expert predators. Utahraptor was one of the biggest at around 6 metres in length.

Extra sharp teeth

DINO FACTS

Name: *Utahraptor*

Meaning: *"Utah's plunderer"*

Size comparison:

Food: *Meat*

Danger rating: *7/10*

Where: *Woodland and floodplains*

Glossary box:

A predator is an animal that hunts other animals to eat.

Dot to Dot!

Join the dots to complete the Utahraptor, then colour it in!

SWIPE!

Agile body

Huge toe claws on back feet

Slashing Claws

Like other dromaeosaurs, Utahraptor had special claws on its back legs that were used to cut and rip its prey. These deadly, killing claws were an impressive 24 centimetres in length!

Body Features

Utahraptor stood upright and walked on its two back legs. With a long, stiff tail and strong but hollow bones, its body was built for speed. Scientists think it also had excellent hearing, which may have helped it sneak up and attack its prey.

DINOSAUR SCALES

Make your very own scaly dinosaur from eggshells.

You will need:

- Card or paper
- a pencil • scissors
- eggshells • pva glue
- a glue brush
- paintbrushes
- paints

Ask an adult to help ⚠

Carefully follow the steps below.

Step 1:

Ask an adult to help you boil a few eggs. Once they have cooled, peel off the shells and carefully break them into small pieces. Set aside.

Step 2:

Draw the outline of your favourite dinosaur on the card or paper. You could use any of the dinosaurs in this book!

Step 3:

Carefully cut it out using the scissors.

Step 4:

Using your glue and brush, stick the broken-up pieces of eggshell to the dinosaur shape to give it scales. Make sure to squash them flat!

Step 5:

When the glue has dried, it should look like this.

Step 6:

Now the dinosaur is ready to paint! Which colours will you choose?

The colourful, hard eggshells on your dinosaur look like scales. Why don't you try making some more dinosaurs and experiment with different textures. Try gluing sand, feathers or cotton wool to your picture. What else could you use?

QUETZALCOATLUS NORTHROPI
The largest pterosaur of all time!

WHOOSH!

11 metre wingspan

Long neck

DINO FACTS

Name: *Quetzalcoatlus Northropi*

Meaning: *Named after the Aztec god, Quetzalcoatl*

Size comparison:

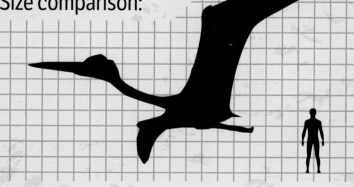

Food: *Meat*

Danger rating: *6/10*

Where: *Skies above floodplains*

Fossil Findings

This species from the Cretaceous period is known from fossils found in Texas and Montana in the USA. Preserved remains of wing bones and skull parts gave scientists clues to its anatomy and movement. It is thought that it walked using its wings and legs.

SWOOP!

Giant Wings

The wingspan of this flying reptile could reach up to 11 metres! It is thought that another Quetzalcoatlus species lived alongside northropi but that it was much smaller, with a wingspan of only 4.5 metres.

Dive and Scoop

Experts believe that this winged giant could glide through the air like a modern-day condor – a large bird of prey. If it spotted a tasty fish below, it probably swooped down to scoop it up in its long, toothless jaws.

Spot the Difference!

MINI Activity

Circle the ten differences between the two pictures.

DINOSAUR SCRAMBLE

The names of these dinosaurs are mixed up. Can you work out the name of each dinosaur and match them to the correct picture?

1

2

3

4

5

NIMMI

_ _ _ _ _

Clue!
You can find all the names in this annual.

SAUSIOPLESUR

_ _ _ _ _ _ _ _ _ _ _

POOLVERCTAIR

_ _ _ _ _ _ _ _ _ _ _

NOODANGUI

_ _ _ _ _ _ _ _ _

COSYTASSRAURU

_ _ _ _ _ _ _ _ _ _ _ _ _

MEMORY GAME

Test your memory with this puzzling prehistoric picture.

Look at the picture for 30 seconds

Try to remember everything.

Cover the picture and answer the questions.

1 How many volcanos?

2 How many trees?

3 What colour are the eggs?

4 How many of these dinosaurs?

5 Circle the dinosaur that is NOT in the picture.

AMAZING MUSEUMS

A museum is a building where important objects, such as dinosaur fossils, are displayed and looked after.

What is a Museum Curator?

Many people work at museums — it takes a lot of work to take care of all the collections! It's the museum curator's job to look after the objects and artifacts by making sure they are organised and stored correctly. They also help with the exhibits and make sure visitors — like you — get the most from their experiences.

It's my job to manage all the collections. I take care of the items and make sure they are shown in the best way so that visitors can learn all about them!

Where can I see dinosaur bones?

All around the world! There are many museums that have dinosaur exhibitions on display — ask an adult to help you research the ones closest to you. Some of the museums have real fossils, while others show replicas (copies) to protect the real ones. Many museums also have life-size reconstructions of dinosaurs to help you see just how impressive they were!

A fossil of a Carnotaurus at the Natural History Museum in Paris, France

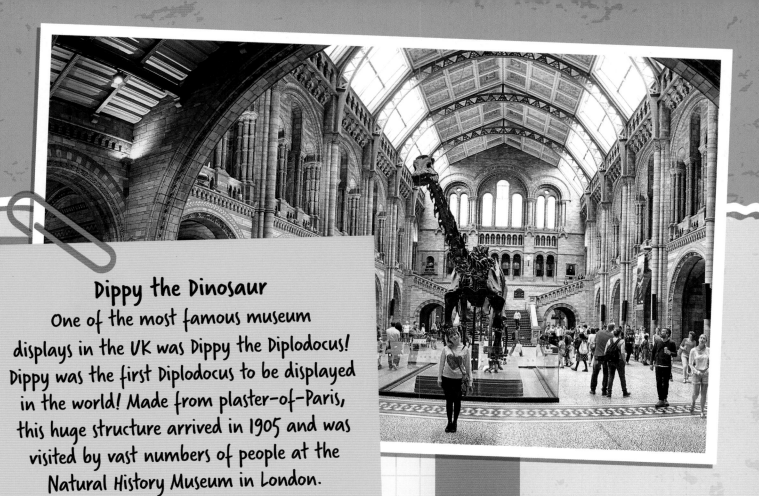

Dippy the Dinosaur

One of the most famous museum displays in the UK was Dippy the Diplodocus! Dippy was the first Diplodocus to be displayed in the world! Made from plaster-of-Paris, this huge structure arrived in 1905 and was visited by vast numbers of people at the Natural History Museum in London.

The skull of a T.rex at the Smithsonian Museum in Washington D.C., USA

Tallest display

This Brachiosaurus skeleton in the Berlin Natural History Museum is the tallest dinosaur on display in the world, standing at 13.3metres!

MINMI

This awesome ankylosaur was covered in bony armour!

Rows of bone

Less than 1 metre in height

Back legs longer than front legs

Thick Skin

Minmi was a fairly small yet heavy dinosaur with excellent protection from predators. Its neck, body and belly all had rows of bony armour. These would have helped protect the dinosaur from being wounded if it was attacked.

DINO FACTS

Name: *Minmi*

Meaning: *Named after Minmi Crossing in Queensland, Australia*

Size comparison:

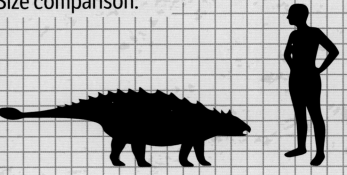

Food: *Plants*

Danger rating: *3/10*

Where: *Floodplains and woodlands*

Australian Discovery

A few Minmi skeletons have been found in Queensland, Australia. One particularly well-preserved example was almost complete, which is quite rare in the world of dinosaur fossil findings!

MUNCH!

Small brain

Body 2–3 metres in length

Fossil Food

One of the Minmi fossil discoveries included a preserved food pellet showing the contents of its stomach when it died. Again, this is a rare find. Inside the pellet, there was lots of plant matter, some fruit and even seeds!

Dinosaur Wordsearch

MINI Activity

Find each dinosaur name in the grid

Words can go left, right, up or down and can bend!

A	S	M	I	N	C	O	L	Y	T
L	U	V	I	M	E	P	H	A	L
L	R	E	A	S	T	S	P	I	E
O	U	L	A	P	R	H	P	N	S
S	A	O	R	N	O	U	O	O	U
A	R	C	A	O	D	S	L	S	R
N	O	I	S	A	U	R	O	A	U
K	T	R	A	R	C	H	I	A	S
Y	P	A	T	S	U	L	A	T	U
L	O	S	A	U	R	A	R	U	R

ALLOSAURUS

ANKYLOSAURUS

ASTRODON

MINMI

PARASAUROLOPHUS

TALARURUS

TARCHIA

TYLOCEPHALE

VELOCIRAPTOR

Answers on page 69.

MATCHING UP!

Draw a line to match the living dinosaur with its fossil.

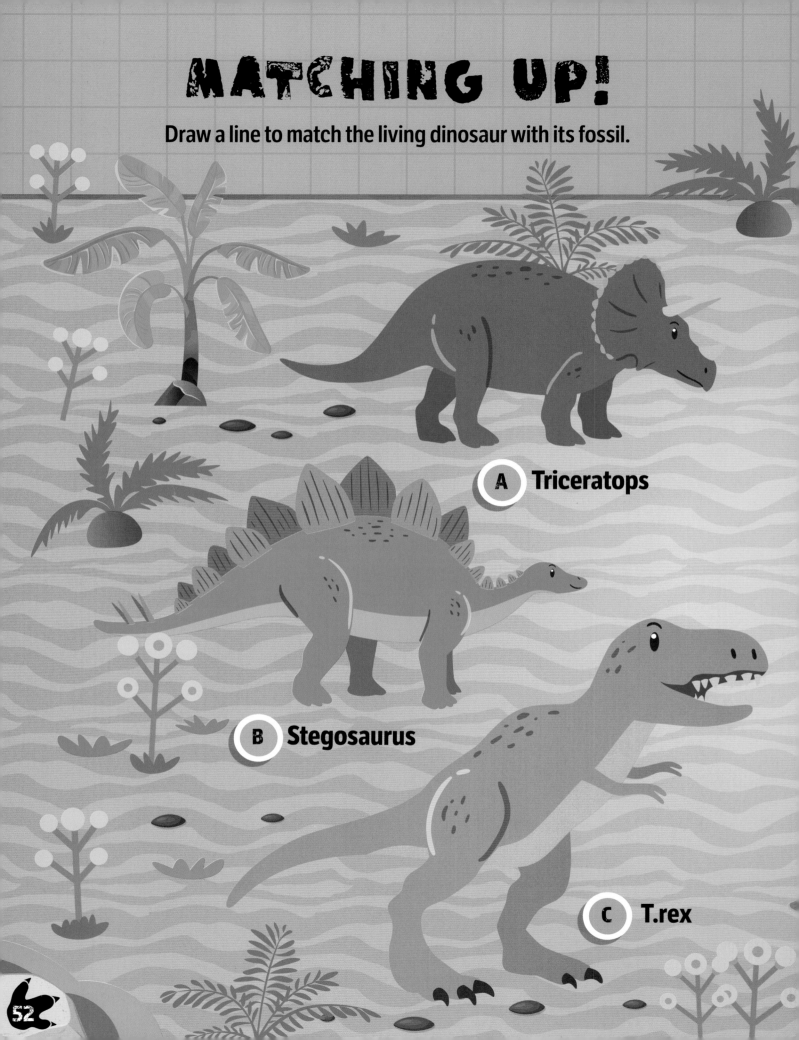

A Triceratops

B Stegosaurus

C T.rex

52

STEGOSAURUS

A slow-moving plant-eater with a spiked tail!

WHACK!

Bony Plates

When this dinosaur was first discovered, scientists thought the plates on its back (scutes) lay flat. This led to it being named "roof lizard." Newer theories suggest that they stood upright instead, with the points facing upwards.

Sharp Spikes

The four impressive tail spikes at the end of Stegosaurus' tail helped fight off predators such as the mighty Allosaurus. The spikes could be almost one metre in length. Now that's a serious weapon!

DINO FACTS

Name: *Stegosaurus*

Meaning: *"Roof lizard"*

Size comparison:

Food: *Plants*

Danger rating: *4/10*

Where: *Forests, floodplains, and grasslands with plenty of vegetation*

Spiked tail

STOMP!

54

Dino Doubles

Two of these spikey dinos are identical. Can you spot which two?

1
2
3
4
5
6
7
8

Tiny Brain

Even though Stegosaurus had a large body, it had a tiny brain! It is quite likely that it wasn't particularly smart – but luckily it didn't need to be. As a plant eater, it wouldn't have needed to work out how to hunt its prey.

Two rows of kite-shaped plates along the back

Up to 9 metres long

Small head and brain

Front legs shorter than back legs

Broad feet

Living in Herds

Fossil finds suggest that Stegosaurus lived, travelled and grazed in groups. These dinosaurs probably walked together, munching on plants that were growing low on the ground, such as grass and ferns.

CHOMP!

Answers on page 69.

HOW TO DRAW

Follow the steps below to draw the picture of the dinosaur on the notepad.

1 Using a pencil, start with some simple ovals and triangles for the head, neck, body and tail.

2 Add circles and long ovals for the top of the front leg and arm.

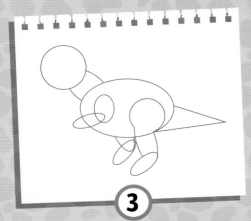

3 Add ovals for both legs and feet.

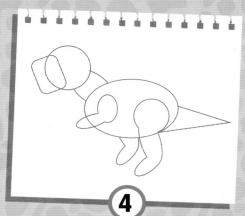

4 Draw a rectangle for the dinosaur's mouth.

5 Add small ovals for the claws and draw a trianglular shape for the mouth.

6 Rub out any overlapping lines to form the head, arms and leg shapes. Add a small circle with a tiny circle within it for the eye and a tiny circle for a nose hole.

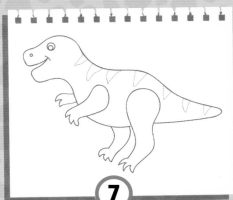

7 When you are happy with your picture draw around the dinosaur shape in pen, erasing any pencil lines you don't need. Add some stripes.

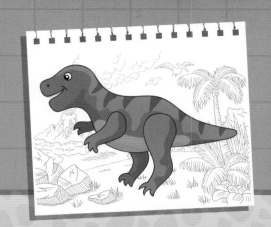

Once you've finished your drawing you can colour in
your dinosaur using the picture at the top as a guide.
Can you add in lots of extra details, such as rocks,
leaves, some clouds and a volcano?

NEW DISCOVERIES

Every year paleontologists make exciting new dinosaur discoveries!

RECENT FINDS

The discovery of a small, armoured dinosaur from the Cretaceous period was announced in 2022. Named Jakapil kaniukara, it is thought to have walked upright on two back legs. Weighing no more than 7 kilograms, it had bony discs down its back for protection from predators.

An almost armless dinosaur species was recently found in Argentina. This new species, Guemesia ochoai, had very small forelimbs that wouldn't have been able to grasp prey. This tells us that it must have relied on other hunting methods — such as a powerful jaw — to overpower prey.

Another dinosaur with small arms, named Meraxes gigas, shows that tiny forelimbs were more common than previously thought. Experts are still trying to work out why certain meat-eating dinosaurs developed this body feature.

Every time a new kind of dinosaur is discovered, scientists learn more about dinosaur science. They find new clues that reveal how different species walked or moved and how and what they ate.

Skull fossils from two new spinosaurs were found on the Isle of Wight, UK in 2013. The names they were given are Ceratosuchops inferodios (meaning "hell heron horned crocodile-face") and Riparovenator milnerae (meaning "riverbank hunter").

Spinosaur

Four small fossils from a meat-eating dinosaur that lived more than 200 million years ago have been recognised as a new species. The dinosaur has been named Pendraig, which means "chief dragon"!

Skin is very rarely preserved; however, a dinosaur leg was found at the Tanis fossil site in North Dakota, USA, with some skin intact. Scientists think the Thescelosaurus could have been killed on the very day the asteroid that wiped out the dinosaurs hit Earth!

COMPSOGNATHUS

A turkey-sized meat eater!

WHIZZ!

Long tail

Small and Speedy!

This little dinosaur dashed around to catch its prey! It is thought that it ate lizards, insects and small mammals, and probably hunted in packs. Its sharp claws and small body may have allowed it to clamber up trees.

Feathery Body

Many scientists believe that Compsognathus may have had a feathery body for warmth – even though no fossils show this to be the case. They think it is possible, however, as some of its closest relatives did have feathers.

DINO FACTS

Name: *Compsognathus*

Meaning: *"Pretty jaw"*

Size comparison:

Food: *Meat*

Danger rating: *5/10*

Where: *Forests, near rivers and lakes*

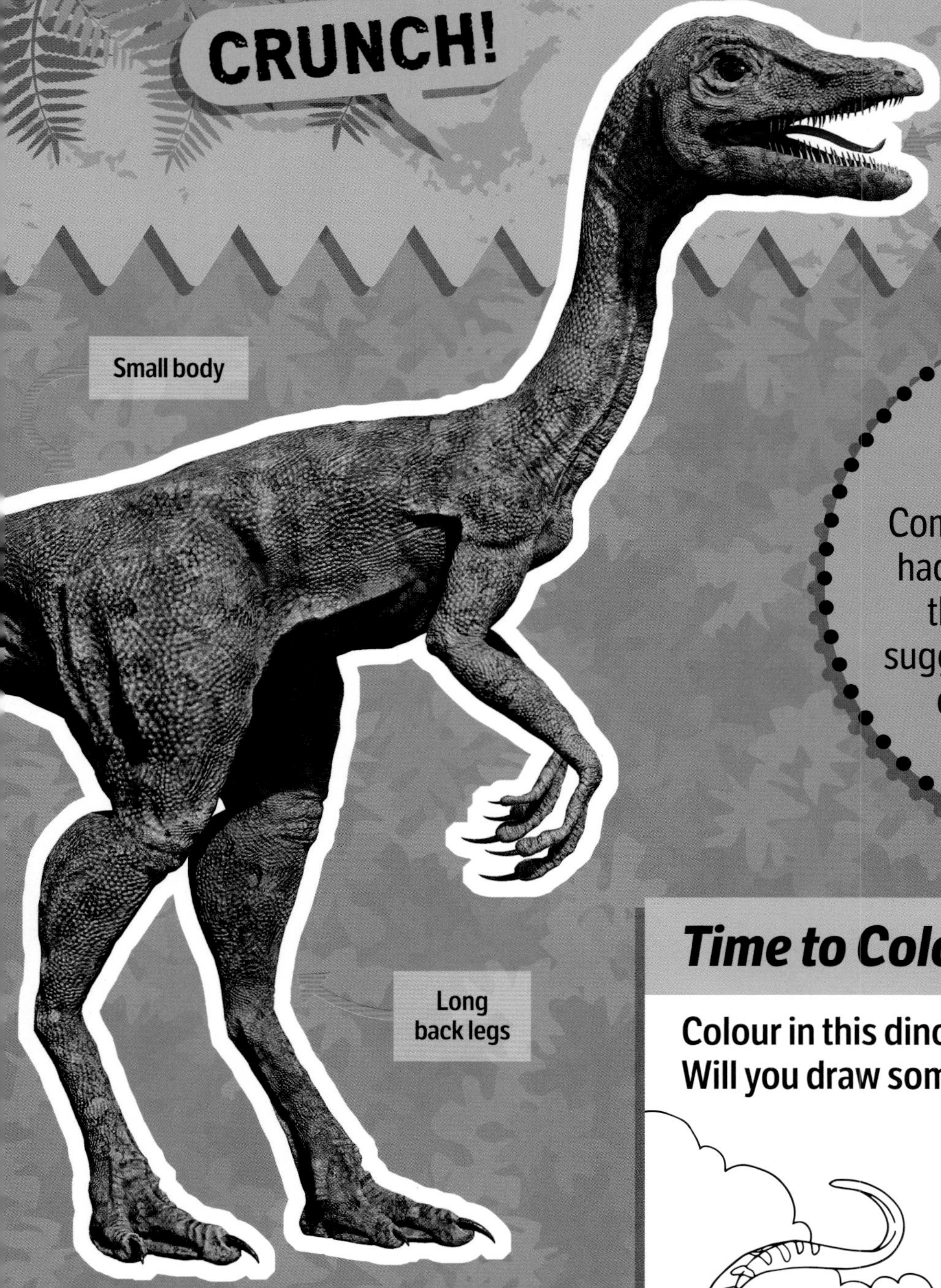

CRUNCH!

Sharp, pointed teeth

Small body

Long back legs

Big Eyes

Fossil remains of Compsognathus show that it had large eye sockets – and therefore big eyes. This suggests that it needed good eyesight to spot – and attack – its prey.

MINI Activity

Time to Colour

Colour in this dinosaur.
Will you draw some feathers on its body?

Glossary box:

A prey animal is one hunted by another animal for food.

WORDSEARCH

There are ten words hidden in the grid below.
Can you find them all? Tick the correct box each time you spot one!

H	S	C	A	L	E	S	Y	E	C	P	Z	X	E
Y	D	R	U	Y	U	A	T	D	O	O	V	Z	K
R	G	M	M	N	B	I	S	D	P	M	E	Z	R
M	F	L	C	O	O	X	P	O	R	I	P	B	Q
T	T	I	I	P	U	C	I	V	O	C	T	S	F
X	H	R	P	Z	H	C	K	L	L	R	D	R	O
A	G	A	E	P	A	Z	E	P	I	O	I	W	S
F	D	O	G	X	S	R	S	D	T	R	N	B	S
B	U	F	N	O	B	K	D	X	E	A	O	F	I
A	F	U	I	K	M	H	K	G	H	P	S	J	L
B	H	F	A	H	W	I	J	B	M	T	A	G	S
W	U	L	J	N	Z	W	Z	T	C	O	U	N	V
M	L	E	X	B	C	U	W	E	J	R	R	L	B
G	R	A	R	U	O	G	V	J	R	I	S	G	D

- [ ] COPROLITE
- [ ] MICRORAPTOR
- [ ] THAGOMIZER
- [ ] T. REX
- [ ] FOSSILS
- [ ] LIZARD
- [ ] FLEA
- [ ] SPIKES
- [ ] DINOSAURS
- [ ] SCALES

62

Answers on page 69.

CROSSWORD

Use your knowledge to solve the clues and fill in the crossword.
All the answers are somewhere in this book!

Hint! Letters are shared when the words cross over each other.

CLUES

ACROSS

2 Animal that hunts another animal for food.

4 Building that displays historical artifacts to the public.

6 Type of space rock beginning with the letter A that crashed into the Earth 66 million years ago, ending the reign of the dinosaurs.

7 Earth's original supercontinent that has split up into smaller continents over time.

DOWN

1 Group of flying reptiles beginning with the letter P.

3 Recently discovered dinosaur, whose name means "chief dragon."

5 Preserved rocky remains of an animal or plant.

Answers on page 69.

VELOCIRAPTOR

The famously feathered dinosaur!

BITE!

DINO FACTS

Name: *Velociraptor*

Meaning: *"Speedy thief"*

Size comparison:

Food: *Meat*

Danger rating: *7/10*

Where: *Deserts*

SLASH!

Sharp teeth

Large brain compared to body size

Feathered Body

It is thought that a Velociraptor's body, arms and tail were covered in feathers – but not for flight. Instead, the feathers may have been used to keep its eggs warm while sitting on its nest – or to impress other Velociraptors.

Super Hunters

The body of a Velociraptor was well suited for catching prey such as lizards and even other small dinosaurs. It had a mouthful of razor-sharp teeth, strong, grasping hands and ferocious claws on its back feet! They hunted – and scavenged – together in packs.

Desert Home

Velociraptor lived during the late Cretaceous period in the deserts of what is now Mongolia in Asia. Amazingly, the hot environment hasn't changed that much since the dinosaurs roamed the sand dunes.

Feathered body and tail

Super Smart

Experts look at brain size compared to body size to try to work out a dinosaur's intelligence, alongside its lifestyle. Velociraptor had a big brain for its size, suggesting it was able to make clever decisions – such as how to hunt for its next meal!

2 metres in length

MINI Activity

Dino Footprints

Study these footprints, can you spot the odd pair out?

1

2

3

4

5

Answers on page 69.

Deadly claw held off the ground

I SPY

How many of each dinosaur can you find? Write the number in the circles.
Then you can colour in the leaves!

Answers on page 69.

TRUE OR FALSE?

Test your memory by answering the questions!

Circle the correct answer!

1	A herbivore eats meat	True	False
2	An asteroid is a space rock	True	False
3	T.rex had the longest front arms of any dinosaur	True	False
4	Coprolites are fossilised feathers	True	False
5	Birds are living descendants of dinosaurs	True	False
6	Dinosaurs died out at the end of the Jurassic Period	True	False
7	Dippy the dinosaur is a famous type of flying reptile	True	False
8	Mary Anning discovered the first-ever complete Plesiosaurus	True	False
9	Compsognathus is thought to have had tiny eyes	True	False
10	The name Velociraptor means "speedy thief"	True	False

Answers on page 69.

ANSWERS

p.11 DINO BABY MATCH UP

p.12 MAZE CHALLENGE

p.21 COUNTING CREATURES

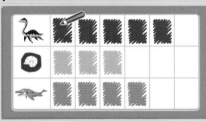

p.25 PREHISTORIC PUZZLE

p.26-27 SPOT THE DIFFERENCE

p.31 FINDING FRIENDS

Line 3 leads the dinosaur to its friends

p.35 SHADOW MATCH

Shadow 2 matches the T.rex

p.45 SPOT THE DIFFERENCE!

p.46 DINOSAUR SCRAMBLE

1. NIMMI (MINMI)
2. NOODANGUI (IGUANODON)
3. POOLVERCTAIR (VELOCIRAPTOR)
4. SAUSIOPLESAR (PLESIOSAURUS)
5. COSYTASSRAURU (STYRACOSAURUS)

p.47 MEMORY GAME

1. Two volcanos
2. Four trees
3. Green eggs
4. Three dinosaurs
5. Stegosaurus is NOT in the picture.

p.51 DINOSAUR WORDSEARCH

A	S	M	I	N	C	O	L	Y	T
L	U	V	I	M	E	P	H	A	L
L	R	E	A	S	T	S	P	I	E
O	U	L	A	P	R	H	P	N	S
S	A	O	R	N	O	U	O	O	U
A	R	C	A	O	D	S	L	S	R
N	O	I	S	A	U	R	O	A	U
K	T	R	A	R	C	H	I	A	S
Y	P	A	T	S	U	L	A	T	U
L	O	S	A	U	R	A	R	U	R

p.52-53 MATCHING UP

A-2, B-1, C-3

p.55 DINO DOUBLES

4 and 7 are identical

p.62 WORDSEARCH

H	S	C	A	L	E	S	Y	E	C	P	Z	X	E
Y	D	R	U	Y	U	A	T	D	O	O	V	Z	K
R	G	M	M	N	B	I	S	D	P	M	E	Z	R
M	F	L	C	O	O	X	P	R	I	P	B	R	Q
T	T	I	I	P	U	C	I	V	O	C	T	F	S
X	H	R	P	Z	H	C	K	L	L	I	D	R	O
A	G	A	E	P	A	Z	E	P	I	O	I	O	S
F	D	O	G	X	S	R	S	D	T	R	N	B	S
B	U	F	N	O	B	K	D	X	E	A	O	I	E
A	F	U	I	K	M	H	K	G	H	P	S	J	L
B	H	F	A	H	W	I	J	B	M	T	A	G	S
W	U	L	J	N	Z	W	Z	T	C	O	U	N	V
M	L	E	X	B	C	U	W	E	J	R	R	L	B
G	R	A	R	U	O	G	V	J	R	I	S	G	D

p.63 CROSSWORD

Across:
2. PREDATOR
4. MUSEUM
6. ASTEROID
7. PANGAEA

Down:
1. PTEROSAURS
3. PENDRAI (PENDRAI)
5. FOSSIL

p.65 DINO FOOTPRINTS

Pair 2 is the odd one out (they are human feet!)

p.66-67 I SPY

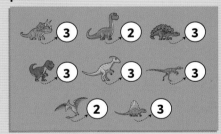

p.68 TRUE OR FALSE?

1. False, a herbivore eats plants
2. True
3. False, T.rex had small front limbs
4. False, it's fossilised poo not feathers.
5. True
6. False, dinosaurs died out at the end of the Cretaceous Period
7. False, Dippy is a Diplodocus!
8. True
9. False, experts believe its large eye sockets meant that its eyes were big.
10. True